Katy Eight

A Numberline Lane book
by
Fiona and Nick Reynolds

"What a lovely day!" thought Katy Eight.

The sun was shining down.

She decided that it was a perfect day for sitting in the garden.

Katy Eight went downstairs and into the kitchen.

She said "Good morning" to her eight pet goldfish, before sitting down to breakfast.

She poured herself some Numberflakes and started to eat them.

After she had finished her flakes, Katy Eight went outside, got out her garden chair and put on her sunglasses.

She was just about to sit down when she noticed that some of her flowers were missing!

She counted them carefully.

"One, two, three, four, five, six... Well that's not right." said Katy Eight, "I know I planted eight pink flowers yesterday. What has happened?"

Just then, Clive Five came walking past.

"Good morning!" said Clive Five.

"What is the matter Katy Eight? Are you alright?"

"I'm not really sure," said Katy Eight. "I planted eight pink flowers yesterday, and now there are only six!"

"Let me help you to count them." said Clive Five.

"One, two, three, four, five, six. You are quite right, there are two missing."

"How did you know that?" asked Katy Eight.

Clive Five explained, "The difference between eight and six is two! So you must have lost two pink flowers".

Katy Eight and Clive Five wondered where the two missing flowers had gone.

Then they heard a chuckle behind them.

Katy Eight and Clive Five turned around to see Linus Minus running down the road with two pink flowers in his hand!

"He has taken away two of my lovely pink flowers," cried Katy Eight, "What am I going to do now?"

"Don't worry," said Clive Five, "I know someone who can help us. We must go and see Gus Plus!"

So off they went to The Add Pad, the home of Gus Plus.

Katy Eight knocked on the door.

"Knock–knock, knock–knock, knock–knock, knock–knock!"

The door opened, and there stood Gus Plus, smiling.

"Welcome, my friends! Do come in."

So Katy Eight and Clive Five went in and told Gus Plus all about the pink flowers, and how Linus Minus had taken two away.

"Don't worry!" said Gus Plus "I can help you."

He took out his Number-wand and said some magic words.

With a flash of light and a puff of smoke the two missing flowers appeared in Katy Eight's hand.

"Thank you Gus Plus! You have solved my problem," said Katy Eight, "With the six flowers in my garden, and these two more, I will have eight flowers again."

Katy Eight and Clive Five walked back to House Number Eight.

She planted her two new flowers.

Then she counted all of them very carefully.

"One, two, three, four, five, six, seven, eight pink flowers!"

She smiled a great big smile.

Clive Five was pleased to see that Katy Eight was happy again.

He said "Goodbye" and left her sitting in her garden enjoying the sunshine.